I0076908

PLANT PROTECTION IN THE PACIFIC 2

EPIDEMIOLOGY OF VANILLA NECROSIS POTYVIRUS IN SELECTED *VANILLA FRAGRANS* PLANTATIONS IN THE KINGDOM OF TONGA.

SEMISI PONE
BSc, MSc (Hons). Auckland.

Copyright © Rainbow Enterprises 2014
Publisher: Rainbow Enterprises 2014

ISBN: 978-1-927308-10-3

All rights reserved. No part of this publication shall be reproduced in any way without prior written consent of the copyright holder and publisher. Rainbow Enterprises is the trade/publishing name of Semisi Pule also known as Semisi Pule Pone.

Distributed by Rainbow Enterprises

Email: semisipone@yahoo.com
rainbowenterprises7@gmail.com

Cover photo: Pinwheel inclusions of Vanilla Necrosis Potyvirus (VNPV) in *Nicotiana benthamiana* leaf tissue thin sections.   Photo courtesy of Professor Michael N Pearson, University of Auckland..

CONTENTS.

LIST OF TABLES.

NOTES ON THE AUTHOR...

Semisi Pule also known as Semisi Pule Pone attended the University of Auckland from 1981 to 1984, graduating with a Bachelor of Science in 1985. He started working for the Ministry of Agriculture, Fisheries and Forests (MAFF) in Tonga as a Plant Pathologist in June, 1985. In 1986, he started the work on vanilla viruses which was also part of a Master of Science programme at the University of Auckland. He graduated with a Master of Science (Honours) in 1989. In 1991, he was appointed the Senior Plant Virologist for MAFF in Tonga. He moved to the University of the South Pacific Agriculture Campus, Apia, Samoa in March, 1992 where he worked as a Fellow in Tissue Culture. In May 1993, he was appointed the Plant Protection Advisor and Co-ordinator for the South Pacific Commission's Plant Protection Service based in Suva, Fiji. He was also appointed to the Food and Agriculture Organisation of the United Nations as a member of the Regional Plant Protection Organisation Technical Consultations and Biosecurity Committee.

The Pacific Plant Protection Organisation was established by a resolution of the South Pacific Conference in Noumea, New Caledonia in 1994. He left the South Pacific Commission in June 1996 and migrated to New Zealand where he was involved with various businesses.

He is now a full-time writer...writing children's stories, novels, poetry, humor and scientific literature.

LIST OF FIGURES.

INTRODUCTION.

This is the second book in this series "Plant Protection in the Pacific". This book will concentrate on the epidemiology of Vanilla Necrosis Potyvirus (VNPV). More detailed discussions and other studies are taken into account since the first detailed study by Pone, 1988.

Vanilla is a very important economic plant to the Kingdom of Tonga . Since 1979 Tonga was an important and significant supplier of dried vanilla beans to the world market, even though Tongan vanilla export is only a small fraction of the total world supply (Table 1). It is a very important source of income for all the vanilla growers of Tonga.

Table 1. Vanilla cured bean export from Tonga 1979-1986.
Source - Project Performance Audit Report, Tonga Development Bank, May, 1988; unpublished.

Year	1979	1980	1981	1982
Metric Tonnes	2	5	9	12
Year	1983	1984	1985	1986
Metric Tonnes	14	12	20	19

Virus-like symptoms of tip, young leaf and stem necrosis which lead to rapid loss of leaves and plant death were observed in the vanilla research plots at the Ministry of Agriculture, Fisheries and Forests, Research Division, Tonga in 1986. Samples were sent to Dr Michael N Pearson, Plant Virologist, of Auckland University for a quick examination and tests of the leaves for plant pathogens. Dr M N Pearson reported back that 3 viruses were found in the samples during electron microscope examination of the sap from the vanilla samples. They were Cymbidium Mosaic Virus (CyMV) and Odontoglossum Ringspot Virus (ORSV), which were common orchid viruses worldwide, and an uncited virus with filamentous, flexuous particles about the same length as a potyvirus.

A survey was subsequently carried out in Tonga to determine how widespread the viruses are in the main vanilla growing islands of Tongatapu and Vava'u.
The viruses were found in all the vanilla plantations visited on Tongatapu and Vava'u Islands. The results of this survey was published by Pearson and Pone (1988).

A research project was launched, funded by the Tonga-German Plant Protection Project* (TGPPP), to try and identify the potyvirus and also examine the relationship between the 3 viruses and the disease symptoms observed on the vanilla plants in the field.

This was a matter of urgency as the viruses posed a threat to the growing vanilla industry in Tonga. It appears there is an epidemic of the viruses threatening Tonga's growing vanilla industry.

---

*-The Tonga-German Plant Protection Project was based at the Ministry of Agriculture, Fisheries and Forests Research Station at Vaini, Tongatapu. It funded the buildings, equipment and scholarships for MAFF staff in Tonga for about 10 years. The Federal Republic of Germany was the donor.

The results of the research project were reported in "**An investigation of 3 virus diseases of** *Vanilla fragrans* **(Salisb.) Ames in the Kingdom of Tonga**" (Pone, 1988). Specific studies were also reported by Pearson et al (1990).

A study on the possibility of using mild virus strain cross-protection to protect the vanilla plants against the severe strain potyvirus was also reported by Liefting, L.,

Pearson, MN and Pone, SP (1992).

The first 3 chapters will discuss and review the evidence of the presence of VNPV in severely affected plants, its pathogenicity and characteristics. Chapter 4 to 7 will discuss the spatial distribution of the VNPV in selected plantations with conclusions discussed in Chapter 8.

The aim of this book is to record the work done on Vanilla Necrosis Potyvirus Virus of *Vanilla fragrans* in Tonga in a form that can be read and used worldwide. Some of the information are less accessible to workers outside New Zealand. For example, the work done by Pone, 1988; which is in the form of a MSc thesis, only available from the Auckland University Biological Science Library and National Library of New Zealand.
Bits of information on vanilla production and developments in Tonga are used as "fillers" to make this presentation more interesting and informative.

CHAPTER 1.   THE EVIDENCE FOR EPIDEMIOLOGICAL STUDIES.

"Epidemiology is the study of how a disease is spread through time and space. The source of disease, modes of transmission, and the factors affecting them are all important in designing control strategies. Thus, before measures to control a virus disease are formulated and applied it is usually necessary to identify the virus causing it, and vectors spreading the virus, as well as the sources of both virus and vectors". (Broadbent, 1964).

Vanilla Necrosis Potyvirus (VNPV) was identified through pathogenicity tests to cause the severe symptoms on vanilla plants (Pone, 1988; Pearson et al, 1991) . They include tip necrosis, chlorotic sunken spots on the leaves which turn necrotic and spread from the shoots down the vine. The necrosis destroys the leaves quickly and vines are left without leaves. Scablike lesions can sometimes be found on leaves and vines. VNPV was associated with symptomatic plants in 89% of 87 symptomatic plants tested (Pone, 1988).

In Koch's postulates* regarding tests of pathogenicity, the organism must be associated with the diseased plants examined. It must also be isolated and inoculated into the same plant species and observed to cause the same symptoms on plants in the field.

---

*- Koch's Postulates is used in Plant Pathology as "the rule" in identifying and proving the pathogenicity of a new pathogen.

VNPV was isolated into *Nicotiana benthamiana* which shows faint veinal chlorosis, *Chenopodium amaranticolor* which shows local chlorotic pinpoint lesions (Pone, 1988). It was also inoculated into young vanilla plants in the laboratory and it caused terminal necrosis of shoots as well as leaf chlorosis and necrosis (Pone, 1988). Characteristics of the VNPV like particle length, pinwheel production, thermal inactivation point, host range, longevity in vitro and dilution end point were recorded (Pone, 1988; Pearson et al, 1990).
A test for determining the presence of the

VNPV in vanilla plants was also necessary to confirm the presence of the virus in vanilla plants. The Enzyme Linked Immuno-Sorbent Assay (ELISA) was chosen because of its sensitivity and ease of use in the field.

Symptomatic vanilla leaves were collected from Tongan vanilla plantations and used in the development of the antiserum for the Double Antibody Sandwich-Immuno-Sorbent Assay (DAS-ELISA) test. The complete results, study and discussions are published in **"Developing an ELISA Test for VNPV in the Kingdom of Tonga"** by Semisi Pone, Rainbow Enterprises (unpublished). A Potyvirus "group test" was later purchased from Sigma USA to be used to confirm the presence of potyviruses in severely symptomatic samples. It was also based on the DAS-ELISA.

DESCRIPTION OF VANILLA NECROSIS POTYVIRUS.

VNPV causes tip necrosis and dieback, chlorotic, necrotic and scablike lesions on leaves and vine and leaf distortion of vanilla (Pearson and Pone, 1988; Pearson *et al*, 1990; Pone, 1988;). It can reduce the growth of new vanilla plants significantly (Pone, 1988). It has flexuous filamentous particles about 776-822 nanometres (nm) long (Pearson *et al*, 1990; Pone, 1988). It is not a stable virus and can only survive in crude sap for 26 hours at room temperature before its infectivity is lost rapidly. At high temperatures, its infectivity is lost at 58°C. In vanilla plants VNPV virus yield was found to be 1.26mg/ml to 2.9mg/ml from 100gms frozen leaf tissue based on A260 values (UV absorption)(Pone, 1988). It is a member of the Potyvirus Y Group (Pearson *et al*, 1990). Vectors include *Aphis gossypii* and possibly other aphids (Pearson *et al*, 1990; Pone, 1988). It can also be transmitted easily, mechanically (Pone, 1988).

The results of the studies confirmed that the virus identified, isolated and described as Vanilla Necrosis Potyvirus is the pathogen causing the vanilla dieback disease. It is possible to use the symptoms of the VNPV on vanilla plants as an indicator of virus presence with some confidence. Although mottled and non-symptomatic plants were tested positive to VNPV using ELISA, a long term study of the epidemiology of the virus within a plantation will show a 100% co-relation between symptom and VNPV presence such as the 5 month mapping and monitoring study of VNPV in Plot 1.

VECTOR

In studies of virus spread it is important to determine how it is spread. Fortunately, for potyviruses, they are well known to be transmitted by aphids in a non-persistent manner. Studies by myself (Pone, 1988) and Dr MN Pearson and others (Pearson et al, 1991) have determined that aphids do probe or feed on vanilla leaves and they do transmit the VNPV between *N. benthamiana* plants.

SOURCES OF VIRUS AND VECTOR

A number of weeds around vanilla plantations showing symptoms of leaf yellowing, mosaic and necrosis were tested for VNPV presence. None were found to be infected. (Pone, 1989. unpublished). It was determined with 100% confidence, from this study, that the only source of the VNPV in Tongan vanilla plantations, are infected vanilla plants and cuttings.

This book will concentrate on examining the evidence of VNPV spread in the field and discussing the results of maps published in an **"An investigation of 3 virus diseases of *Vanilla fragrans* (Salisb.) Ames in the Kingdom of Tonga" (Pone, 1988)**.

Conclusions will also be drawn as to the significance of the studies to field control of the VNPV in vanilla plantations, not only in Tonga, but in other countries were VNPV can be found.

CHAPTER 2. EFFECT OF CyMV AND ORSV ON VANILLA PLANTS.

In order to clearly discuss the effect and epidemiology of the Vanilla Necrosis Potyvirus (VNPV), it is important to take into account the effect and possible harm due to CyMV and ORSV. There has been no study dedicated to CyMV and ORSV effects on vanilla in Tonga, but the evidence collected so far suggest they do have a harmful effect like mottling of leaves and perhaps stunting but they are not readily observable in the field. They do not appear to be as harmful as VNPV (Pone, 1988; Pearson and Pone, 1988). Vanilla plants exhibiting mottled leaves still produce a large amount of beans and is not the target of the disease control and epidemiology surveys.

CyMV - Cymbidium Mosaic Virus

CyMV causes a mosaic and necrosis in several genera of orchids (Jensen, 1951; Jensen and Gold, 1955; Kado and Jensen,

1964) and can reduce the growth of Cymbidium orchids significantly (Pearson and Cole, 1986). It has flexuous filamentous particles about 475 nanometres (nm) long and 13 nm wide (Francki, 1970). It is a fairly stable virus which can survive in crude sap, for up to 1 week, at room temperature and up to 70°C for 10 minutes. It multiplies in orchids to very high concentrations, yielding about 360 mg per kilogram of leaves. It contains 6% RNA and is a member of the Potato Virus X group and has no known vector. It can be transmitted easily, mechanically (Francki, 1970).

It was found in 63% of vanilla leaves with severe symptoms, 50% of vanilla leaves with mottles and 23% of leaves with no symptoms in the survey (Pearson and Pone, 1988) (Table 2).

Table 2. Percentage (%) of incidence of viruses in leaf samples of *V. fragrans* from Tonga (30 samples). (Pearson and Pone, 1988).

Symptom	VNPV	CyMV	ORSV	Not detected
A	67	63	27	13
B	3	50	20	40
C	0	23	3	77

Note: A = severe symptoms, B = mottle symptoms, C = no symptoms.

CyMV was also found in 71% of 87 symptomatic samples, 55% of 61 mottled samples and 8% of 105 non-symptomatic plants surveyed (Pone, 1988) (Table 3). It should be noted that although CyMV was found to be present with both VNPV and ORSV in vanilla plants with severe symptoms, VNPV by itself can induce the severe symptoms in the laboratory pathogenicity tests.

Table 3. Incidence of viruses, singly and in combinations, in leaf samples from *V. fragrans* plantations in Tonga (Pone, 1988).

Symptom type	A	B	C
Number of plants tested	87	61	105
VNPV only	14	2	1
CyMV only	2	2	3
ORSV only	1	15	45
VNPV+CyMV	16	1	0
VNPV+ORSV	5	1	0
CyMV+ORSV	2	24	6
VNPV+CyMV+ORSV	42	9	0
Not detected	5	7	50

Note: A = severe symptoms, B = mottle symptoms, C = no symptoms.

In pathogenicity tests 3 of 5 plants inoculated with CyMV infected vanilla inoculum, tested positive to CyMV after several weeks   using ELISA and ISEM*, but plants did not show any symptoms.

* - ISEM refers to Immuno-Sorbent Electron Microscopy or the use of the antiserum to specific viruses to trap virus particles on the Electron Microscope (EM) grids before observation under the EM. The result is much better than using grids with no antiserum.

The effect of CyMV on orchids in general is noted to be severe. Both severe symptoms of mosaic and necrosis. However, the evidence from the surveys and pathogenicity studies so far does not suggest that CyMV causes the vanilla dieback by itself. It is present in a large number of vanilla plants with severe symptoms, in combination with VNPV, but it is also present in a large number of vanilla plants with mottles, in combination with ORSV, and in a small number of plants with no symptoms at all (Table 3).

There were 2 plants with severe symptoms (Table 3) which were positive to CyMV only. This can be explained by the 1. large range of symptoms used, which included chlorotic lesions, necrotic lesions, leaf distortion and scablike lesions (It may be, under certain environmental conditions and vanilla plant susceptibility that CyMV by itself can induce "potyviruslike" symptoms).   2. presence of mild VNPV strains (Liefting *et al*,1992) that may not be serologically related and serologically non-related vanilla potyviruses (Wisler *et al*, 1987a,b). However, the number is so small, and the evidence does not support it, it was not worth further investigation.

ORSV - Odontoglossum Ringspot Virus.

ORSV has rod shaped particles about 300 nm long and 18 nm wide infecting a wide range of orchids systemically with no apparent symptoms (Paul, 1975). However, it causes ringspots symptoms in Odontoglossum grande (Jensen and Gold, 1951) and hence the name. ORSV can cause significant reductions in the growth of *Cymbidium* orchids (Pearson and Cole, 1986). It is a very stable virus surviving at room temperature for over 10 years and 10 minutes at 90°C (Edwardson and Zettler, 1986). It multiplies to very high concentrations in *Nicotiana tabacum* cv *Samsum*, yielding up to 1 gram virus per kilogram of leaf tissue. ORSV particles contain 5% RNA and is a member of the Tobamo Virus Group. No vector is known but ORSV is very contagious mechanically.

ORSV was found in 27% of vanilla leaves with severe symptoms, 20% of vanilla leaves of mottling and 3% of vanilla leaves with no symptoms (Pearson and Pone, 1988) (Table 2).

This survey did not find particles in 13% of leaves with severe symptoms using Electron Microscopy (EM). In EM and ELISA tested samples 40% of samples with mottles tested negative to CyMV or ORSV. Also 77 percent of leaves with not symptoms tested negative to all 3 viruses using EM and ELISA (CyMV and ORSV only) (Pearson and Pone, 1988) (Table 2).

CHAPTER 3. VANILLA NECROSIS POTYVIRUS

COMMENTS ON POTYVIRUSES OF ORCHIDS.

Several potyviruses have been reported to infect Orchids. Leseman and Koenig (1985) found a strain of Bean Yellow Mosaic Virus (BYMV) to infect 20 species of *Masdevallia* exported from the United States to Germany. Infected plants exhibited symptoms of chlorotic streak mosaic on young leaves and surface irregularities on older leaves. Some infected plants did not show symptoms at all. The virus particle length was measured to be 745nm long and parts of pinwheels were also found in leaf thin sections. The virus induced severe necrosis in inoculated *Nicotiana clevelandii* and local lesions on *Chenopodium quinoa*.

Another potyvirus was found on *Orchis militaris*, exhibiting symptoms of leaf mosaic and regrowth reduction. Particle length was 745 nm and parts of pinwheels were also found in leaf thin sections. The virus causes a systemic infection of *N. clevelandii* and local lesions on *C. quinoa*.

An unidentified potyvirus measuring 774 nm and also induces pinwheels was found in *Cypripedium calceolus*. Symptoms of infected plants included chlorotic streaks which turn necrotic and leaf deformation. The virus has not been transmitted experimentally. All these viruses are considered potentially harmful to commercial orchids (Leseman and Vetten, 1985).

A potyvirus was also reported to infect *Vanilla tahitiensis* in French Polynesia (Wisler et al, 1987a, b). It causes a mosaic and distortion of the leaves but does not appear to be severe as VNPV on *Vanilla fragrans* in Tonga.

VANILLA NECROSIS POTYVIRUS CHARACTERISTICS.

"For new viruses, a description of its properties may be required, therefore an investigation of its characteristics is essential. When an unknown virus is investigated, a preliminary characterisation often starts with the determination of particle morphology. This is essential especially for rod shaped and filamentous viruses, which can be identified as belonging to a group from particle size only" (Hamilton *et al*, 1981).

Particle Size

VNPV particle length were measured by taking photos of the virus particles with an Electron Microscope at 10,000x magnification and measured with an Apple Graphics Tablet. Particles from vanilla sap were 776 nm mean length. Particles from partially purified preparations were 822 nm mean length. The differences can be attributed to larger particles being easier to collect at high centrifuge speeds. Both particle lengths are within the 680-900 nm range for the Potyvirus Group. (Pone, 1988).

Particle length of the sap sample had a range from 684 nm to 870 nm which are within the 680 to 900 nm potyvirus particle length as ruled by the International Committee for the Taxonomy of Viruses (I.C.T.V.) (Matthews, 1982).

UV absorption

"The ultra-violet absorption (220-300 nm) of potyviruses are characteristic for the group with a maximum at 260-262 nm and minimum values at 240-246 nm. The A260/280 ratio is between 1.14-1.25" (Hollings and Brunt, 1981a).

Three partially purified VNPV preparations were made from symptomatic vanilla leaves. The A260/280 ratio were 1.17, 1.16 and 1.15. These readings confirm that the virus particles seen on the Electron Microscope and measured in the particle measurement exercise were indeed potyvirus particles.

Virus yield

VNPV virus particle yield from the purification of VNPV was estimated to be in the range of 1.26mg/ml to 2.9 mg/ml from 100 gms of frozen symptomatic vanilla leaf tissue, based on A260 value (Pone, 1988).

Host Range

"Host range refer to the range of hosts infected by the virus investigated. It is usually characteristic of some viruses and has been used as a guide to virus identification" (Hamilton *et al*, 1981).

Species infected with VNPV infected vanilla leaves were *Chenopodium amaranticolor*, which produced local chlorotic, pinpoint lesions and *Nicotiana benthamiana,* which produced a faint systemic veinal chlorosis (Pone, 1988).

Species infected by VNPV to date include *Chenopodium quinoa, C. amaranticolor, Nicotiana benthamiana* and *N. clevelandii.*

C. quinoa, C. amaranticolor and *N.clevelandii* have been infected with more than 20 potyviruses and seem to be a characteristic of potyviruses (Pone, 1988;

Pearson *et al*, 1990).

The antiserum produced to VNPV was tested against 23 other common potyviruses of monocots from the reference list of the Institute of Horticultural Research, Littlehampton, U.K. using ISEM. All results were negative. There is no indication that VNPV is serologically related to other common potyviruses of monocots (Pearson *et al*, 1990).

Thermal Inactivation Point (TIP).

C. amaranticolor plant leaves were infected with VNPV from infected
N. benthamiana plants, using sap dipped in water at temperatures at 5° intervals from 40°C to 70°C. The control was left at room temperature (approx. 25°C). The number of lesions on the leaves were counted after 12 days. Eight lesions were produced in the control, 40°C and 45°C, but gradually decreased. The infectivity was reduced by increasing temperature and was lost after 58°C (Pone, 1988). The number of lesions

increased slightly at 40°C from 25°C (control). It may be that VNPV becomes more infective at warmer temperatures.

"The VNPV seem to be infective at warm temperatures but gradually loses infectivity at temperatures higher than 40°C".

Longevity *in vitro* (LIV).

The number of VNPV lesions produced on *C. amaranticolor* leaves infected with *N.benthamiana* sap decreased with time. At 0 hours there were 27 lesions produced per leaf, after 2 hours it was reduced to 7. It appears the infectivity is greatly reduced, but not lost,   in the first 2 hours. At 30 hours less than 1 lesion was produced up to 96 hours (Pone, 1988).

'It appears that if an aphid picks up the virus from infected plants the infectivity is greatly reduced after 2 hours although it may remain infective up to 96 hours'.

Dilution End Point (DEP).

The number of lesions produced on *C. amaranticolor* leaves decreased with higher dilutions and infectivity was lost at 1/50,000. The number of lesions for this test was much higher with 178 lesions in 1/5 dilution and less than 1 at 1/50,000 (Pone, 1988). It shows there is a lot of variability in the concentration of VNPV in individual *N.benthamiana* plants. This was a different source plant from the one used for the TIP and LIV tests.

'**Virus transmission by aphids may depend on various factors like temperature, elapsed time from the last probe as well as the concentration of VNPV in the source plant**'.

APHID TRANSMISSION

Aphis gossypii transmitted VNPV between *N.benthamiana* plants (Pearson *et al*, 1990). *Aphis gossypii* was also used to transmit VNPV between vanilla plants in a 3 month study in Tonga. Plants tested negative after the 3 months, using ELISA (Pone, 1988).

Wisler *et al* (1987) transmitted the potyvirus infecting *Vanilla tahitiensis* to *Vanilla pompona* using aphids, but the symptoms took 3-8 months to develop.

Stechmann (pers.comm) observed aphids to feed on vanilla leaves in the laboratory. A common grass aphid in Tonga, *Rhopalosiphum maidis,* was trapped on vanilla leaves. *Aphis gossypii* and *Pentalonia nigronervosa*, the banana aphid, were also found in yellow traps in vanilla plantations in Tonga.

'It is possible that VNPV is being transmitted by aphids between *Vanilla fragrans* in Tonga because they have been observed to feed on vanilla leaves as well as present in *Vanilla fragrans* plantations. VNPV has been shown to be transmitted by *A. gossypii* between *N.benthamiana* plants in the laboratory'.

MECHANICAL TRANSMISSION

VNPV was transmitted from infected vanilla sap to healthy vanilla shoots by rubbing the leaves with carborundum powder and grafting infected wedges into healthy vanilla cuttings. It was also transmitted mechanically from infected vanilla leaves to *N.benthamiana,* plants. VNPV was also transmitted mechanically from *N.benthamiana* plants to *C.amaranticolor* plants.

STUNTED GROWTH

It took 27-33 days for healthy vanilla cuttings in the greenhouse to generate new shoots while the VNPV infected cuttings took 50-113 days to grow new shoots (Pone, 1988).

'VNPV is transmitted by cuttings for new plantations. It also delays new growth and stunts the vanilla plant and new shoots'.

Vanilla rehab in Vava'u.....Queen is sharing the latest sustainable farming techniques; assisting growers to cure their own beans and helping them achieve Organic Certification for their plantations. In addition, growers are being offered a sustainable direct supply partnership with Queen which ensures ongoing future income. Ian Jones, who represents Queen Fine Foods in Vava'u, said that His Majesty's visit to the plantation was very well received by the growers. King Tupou VI could see first-hand all the work that had gone into restoring the vanilla and encouraged the growers to work hard to rehabilitate and maintain their vanilla....online press release....scoop.co.nz

CHAPTER 4. SPATIAL DISTRIBUTION OF VANILLA NECROSIS POTYVIRUS IN SELECTED VANILLA PLANTATIONS.

" The pattern of distribution of virus infected plants in the field can be used to study the spread of the virus infections and therefore become useful in designing control strategies. The most convenient method to study the spread of a disease is to study the distribution of diseased plants in the field or orchard without a knowledge of their history. On average, disease plants are often found to clump around other formerly infected plants (van der Plank, 1960)".

Three vanilla plantations were selected for this study from Tongatapu and Vava'u. Two plantations were selected from Tongatapu growers and one from the Vava'u MAFF Research Station. All plantations were mapped to check the spread of the VNPV between plants.

The two plantations on Tongatapu were mapped and monitored to check the spread of the VNPV between plants every month for 5 months. The MAFF plot in Vava'u was only mapped once.

The second plantation, on Tongatapu, was selected because VNPV was reported to spread very quickly prior to the warm months from September to December, when the mapping was done. Unlike the other 2 plantations which had minimal weeds and aphid numbers, this plantation had a large number of the weeds of the species *Emilia sonchifolia* and *Sonchus oleraceous* which were full of aphids during the time of mapping and monitoring.

"Viruses do not spread on their own, but are carried by vectors. The vector for potyviruses are usually aphids transmitting the virus in a non-persistent manner (Hollings and Brunt, 1981a; Matthews 1982)".

"The spatial pattern of infected plants will be clustered or clumped if the infection of a given plant increases the probability of another plant nearby being infected. In short, with clustering, infected plants are, on average, grouped together and healthy plants are grouped. The method for determining the spatial pattern depends on

collecting of several independent samples or dividing a field into contiguous quadrats (Madden and Campbell, 1986)".

"The pattern of distribution of the population of VNPV infected plants can then be determined by comparing the variance (s^2) and the arithmetic mean (u) of the quadrats in each plot. If s^2 = u the distribution is that of the Poisson series or random. However, when s^2 > u the distribution is of a negative binomial or contagious (clustered) (Elliott, 1977)".

"A test for significance in this case is given by $X^2 = s^2$ (n-1)/u. When the X^2 value is greater than the corresponding value in a chi-square table with n-1 degrees of freedom and significance level P, one rejects the Poisson distribution in favour of clustering (Madden and Campbell, 1986)".

Tonga has been chosen for this programme due to the high quality of their vanilla beans, which possess a similar flavour and aroma profile to those from the east-African region of Madagascar and the surrounding countries. Vanilla once grew in the region, predominantly in the 80's and 90's. However, due to sustained poor prices and a range of growing and trading factors, the industry effectively collapsed. Queen's rehabilitation programme addresses these factors to provide long term, sustainable, viable farming methods in the ever-changing vanilla industry. Based in Brisbane, Queen Fine Foods is a family-owned business that has been making fine baking products since 1897 and is now one of New Zealand's leading distributors of home baking and decorating products. Queen's vanilla range includes their award-winning Vanilla Paste, Vanilla Extract, Vanilla Beans and Vanilla Bean Dusting Sugar......online press release.....scoop.co.nz

FIGURE 1. VANILLA PLOT 1.

A LOCAL GROWER ON TONGATAPU.
Note. Infected vanilla plants are the "dark spots" on the map. (Gower: Tongotea of Ha'akame).

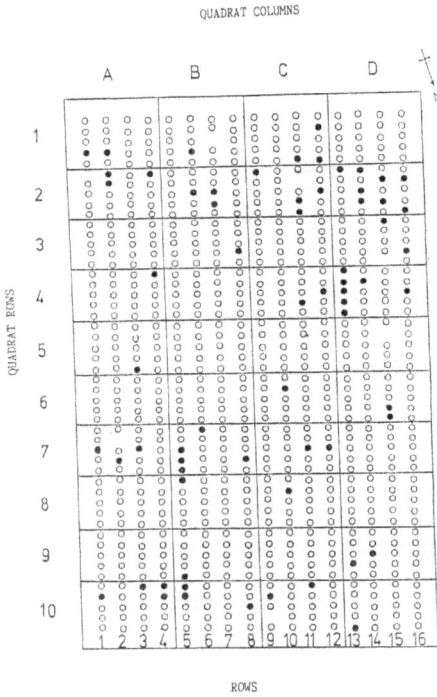

FIGURE 2. VANILLA PLOT 2.

A LOCAL GROWER ON TONGATAPU.
Note. Infected vanilla plants are the "dark spots" on the map. (Grower: Henry Raas of Ha'asini).

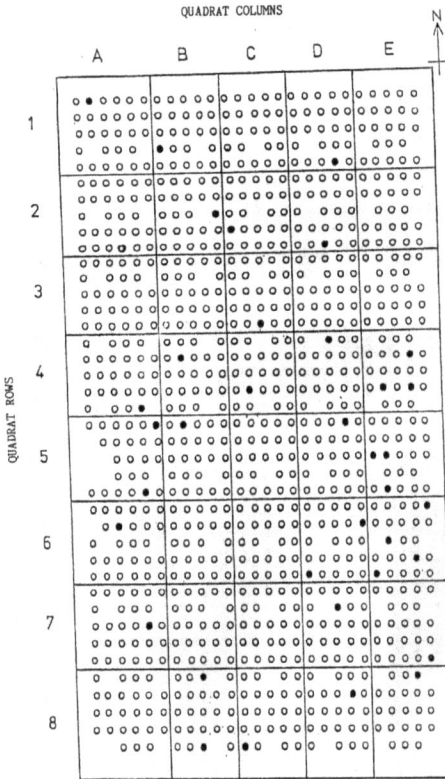

FIGURE 3. VANILLA PLOT 3.

**VAVA'U MAFF RESEARCH STATION
EXPERIMENTAL PLOT. Note. Infected vanilla
plants are the "dark spots" on the map.**

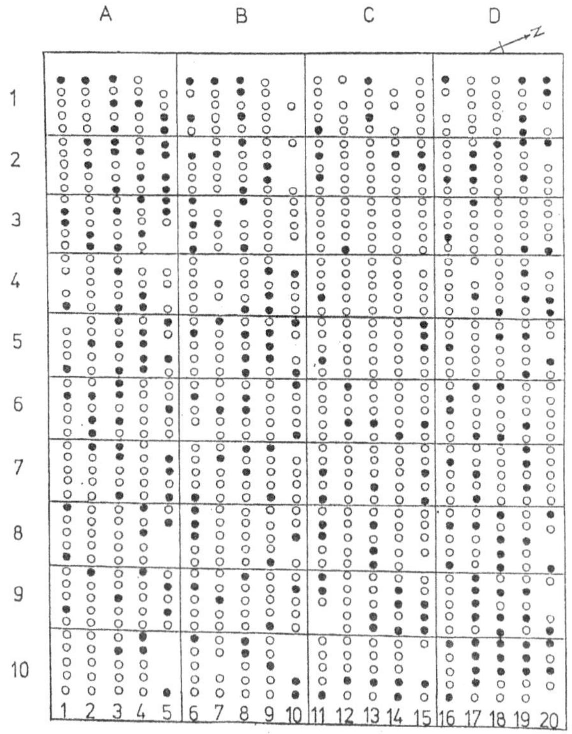

Table 4. The number of symptomatic vanilla plants per quadrat in Plot 1.

			Quadrat columns		
		A	B	C	D
Quadrat	1	2	1	3	0
Rows	2	3	3	4	8
	3	0	1	0	2
	4	1	0	2	7
	5	1	0	0	0
	6	0	0	1	2
	7	3	5	2	0
	8	0	1	1	0
	9	0	1	0	2
	10	4	3	2	1

The Heilala Vanilla Story...*a story of a Tongan aid project that blossomed into a business which creates a range of 100% Pure Vanilla products for foodies around the globe...*
The Boggiss and Ross family started Heilala Vanilla in 2002 and still own and operate the business today. Before 2002, John was a retired dairy farmer, Jennifer was an accountant and Garth worked in IT. They worked a dormant piece of land in Utungake, Tonga gifted to them by the local village. Little did they know at the time that the piece of land was destined for something great. John and Garth put to practice their horticultural know how to kick start the plantation by researching countries around the world that grew vanilla in the narrow band 20 degrees on each side of the equator.

Table 5. The number of symptomatic vanilla plants per quadrat in Plot 2.

| | | Quadrat columns | | | | |
		A	B	C	D	E
	1	1	1	0	1	0
	2	0	1	1	1	0
Quadrat	3	0	0	1	0	0
Rows	4	1	1	1	1	3
	5	2	1	0	1	3
	6	1	0	0	2	4
	7	1	0	0	1	1
	8	0	2	1	1	1

The Heilala Vanilla Story....continued

The plan was to help to provide the locals with employment and hope that the demand for vanilla blossomed.It then took three years to develop and nurture the vines through the on-going art of careful training, weeding and looping, all while ensuring organic sustainable farming was being practised. John who was once a frequent holiday maker to Tonga is now virtually a local spending up to six months a year at the Tongan plantation.In 2005 the first 40kg harvest of vanilla beans were ready. Time passed, the plantation went from strength to strength harvesting a healthy two tonne in 2010.

Table 6 . The number of symptomatic vanilla plants per quadrat in Plot 3.

		Quadrat columns			
		A	B	C	D
	1	8	6	3	6
	2	11	6	5	8
Quadrat	3	10	6	1	4
Rows	4	5	5	1	6
	5	11	10	4	5
	6	8	6	5	7
	7	7	6	5	6
	8	5	5	5	10
	9	6	5	10	10
	10	4	6	6	14

The Heilala Vanilla Story....continued

All the tender love and care has resulted in the richest grade of Vanilla in the Asia Pacific region with its distinctive aroma, shine and plumpness coveted by chefs all around the world. An annual crop is brought back from Tonga to the company's base in Tauranga, New Zealand. Heilala Vanilla is then packaged for each order; the Pure Extract and Vanilla Paste, Syrup and vanilla bean sugar are manufactured, and dispatched to chefs, gourmet food manufacturers and a selection of specialty retail outlets.

Table 7. The variances and the means of plots 1, 2 and 3 for the two quadrat sizes used.

	Quadrat size				
	Large			Small	
Plot	mean	variance		mean	variance
1	3.3	6.5		1.675	3.76
2	1.8	2.7		0.875	0.8814
3	12.8	22.48		6.45	7.06

The Heilala Vanilla Story.....concluded.

Several years have passed and the plantation has matured, but the research and development of more exciting 100% pure vanilla ideas continue. The practice of true sustainability with the local village also continues and has enabled resources for education and infrastructure, which the community otherwise may not have had. It is recognised by the local Agriculture Ministry as a true example of a Pacific partnership in practice something that is rather special to us.

Note from the author....Heilala Vanilla is a shining example of what can be done with Tonga's vanilla and I selected this story from the Heilala Vanilla website.

45

Table 8. The chi-square statistic for plot 1, 2 and 3 for the two quadrat sizes used at P = 0.10, 0.05, 0.025 levels of significance.

	Quadrat size				
Plot	Large		Small		
	X^2 cal.	X^2 tab.(19 d.f.). $P=0.10$ 0.05 0.025 27.2 30.14 32.85		X^2	X^2 tab.(39 d.f.). $P=0.10$ 0.05 0.025 51.8 55.7 73.4
1.	37.42		87.54		
2.	28.5		39.28		
3.	33.37		42.7		

The Story of Tongan Vanilla.......from their website....another success story...

Tonga Vanilla has several acres of vanilla (*planifolia*) growing in plantation and in shade houses, here on the shady cool western slopes of the South Pacific island of `Eua - just a few minutes flight from Tongatapu.
We produce fine vanilla for individuals around the world. Our products are available in a range of quantities, carefully packaged and sealed to ensure the freshness and quality of our rich product. This family-run business has been in operation for eight years, and we look forward to hearing from you, and sharing with you some of our fine Tonga Vanilla.

CHAPTER 5. LINEAR DISTRIBUTION OF VNPV INFECTED VANILLA PLANTS IN THE 3 PLOTS.

Distributions of VNPV infected plants along the vanilla rows.

"Mechanical transmission of viruses in the field by natural mechanical damage to the plant tissue is relatively rare, and probably of very minor economic importance. A more common means of mechanical transmission in the field is through normal horticultural practices. For example, Tobacco Mosaic Virus may be transmitted in tomato crops by contaminated hands, clothing and tools. Many other viruses may be transmitted by unsterilised tools during pruning procedures and when cuttings are taken (Walkey, 1985)".

Vanilla fragrans are planted in rows with a spacing of 1.5 m between plants and 2.5 m between rows giving rise to approximately 2200-2500 plants per hectare. *Vanilla* growers normally work along a row. Since VNPV can be mechanically transmitted, it is highly likely that it is spread along the rows during cultural practices such as looping, flower initiation and taking of cuttings.

"If this assumption is true (mechanical transmission), then the actual number of consecutive infected plants along a row will be greater than the number of consecutive infected plants expected by a chance infection. Therefore, a test for the spread of virus infection is a test for the aggregation of virus infected plants along a row (van der Plank, 1960)".

The Doublet Analysis

There were 2 methods used to analyse the spread of the VNPV along vanilla rows. The first one is described by van der Plank, 1960. He proposed a test for the spread of viruses along the rows called a "Doublet Analysis".

"A doublet consist of 2 adjacent infected plants, three adjacent infected plants is considered as 2 doublets, 4 adjacent infected plants are considered 3 doublets and so on (van der Plank, 1960)".

The Null Hypothesis evaluated is that the ordered sequence is random; the alternative hypothesis is that the sequence is clustered (Madden and Campbell, 1986).

If n plants are investigated in a sequence, and of these u are diseased, then the expected number of doublets d is given by $d = u(u-1)/n$ (van der Plank, 1960).

Madden and Campbell, 1986 proposed an "improved" version of the doublet analysis as $E(D) = m(m-1)/N$ where $E(D)$ is the expected number of doublets, m is the number of infected plants in a row, and N is the total number of plants in that row. The standard deviation $s(D)$ can be calculated by
$s(D) = \{[m(m-1)/N](1-2)/N\}^{\frac{1}{2}}$.

For rows with $N > 20$ a test of randomness is given by $Z = (D-E(D)/s(D)$.

If $Z > 1.64$, one rejects the null hypothesis of randomness in favour of clustering $(P = 0.05)$.

The Ordinary Run Analysis.

A second method used to analyse the spread of the VNPV along vanilla rows is described by Madden and Campbell.

"Madden et al, 1982 proposed the "ordinary run analysis" where in sequences of diseased and healthy plants in a row, a run is defined as a consecutive clump of either healthy or infected plants. Using this system, the expected number of runs E(U) is given by;

$$E(U) = 1 + 2m(N-m)/N$$

With standard deviation;

$$s(U) = \{[2m(N-m)-N]/[N2\ (N-1)]\}^{1/2}$$

A test of randomness (N >20) is given by;

$$Z(U) = [U + 0.5 - E(D)]/s(U)$$

If -Z(U) > 1.64, then one rejects the null hypothesis of randomness in favour of clustering.

THE ANALYSIS

The doublet and ordinary run analysis were applied to the mapping and monitoring results of plots 1 and 3 where significant numbers of vanilla plants were infected. It is noted that the Plot number 2 which had a small number of infected plants, and the virus was not spreading, had bare ground with no aphid hosts plants near the vanilla plants during the assessment.

Table 9. Doublet Analysis of the distribution of VNPV infected plants along vanilla rows in Plot 1.

Rows	D	m	N	E(D)	s(D)	Z
1	0	3	49	0.122	0.342	-0.356
2	1	4	49	0.244	0.48	1.57
3	0	3	50	0.12	0.339	-0.353
4	1	4	50	0.24	0.48	1.585
5	5	7	49	0.857	0.9	4.6*
6	0	3	50	0.12	0.339	-0.353
7	1	2	50	0.04	0.196	4.89*
8	0	3	50	0.12	0.339	-0.353
9	0	2	49	0.04	0.196	-0.558
10	0	2	50	0.04	0.196	-0.558
11	1	6	49	0.6122	0.766	0.506
12	0	5	50	0.4	0.619	-0.646
13	4	8	50	1.12	1.037	2.77*
14	1	5	50	0.4	0.619	0.969
15	1	5	48	0.416	0.63	0.926
16	0	4	50	0.24	0.48	-0.5

* - significant at P = 0.05

ISLANDS BUSINESS....Tonga's vanilla industry is looking promising again as local growers and exporters clinch new deals likely to lift output over the next year.   This comes as the island kingdom enters a very difficult phase economically and any progress in local industries is an important ...contribution to its recovery. "The vanilla industry used to be very strong many years ago but over time it has decreased with the fluctuating international process," Ian Jones, director of Vava'u-based virgin coconut oil exporter Taste of Tonga told ISLANDS BUSINESS. Vava'u's warm tropical climate and generally fertile soil makes it an ideal location for vanilla cultivation and it is Tonga's vanilla growing center. But Tonga's vanilla heydays are gone.

"Only 105 of the plantations are currently producing vanilla beans. Most are neglected," said Jones. He plays a key role in a new partnership between local vanilla growers and Queen Fine Foods, an Australian-based family-owned business specialising in the development of vanilla products and their distribution in Australia and New Zealand.

Queen's investment, significant by Tonga's vanilla industry scale, will see the development of the Queen Vanilla Curing Certification Course (QVCCC) to help grower members of the scheme develop sustainable farming practices. It will also provide farmers with the critical curing facilities and farmers' education in vanilla curing. Jones, a director of Vava'u-based Taste of Tonga and principally a virgin coconut oil exporter, will help this vanilla project as Queen's man on the ground in Vava'u to roll out its vanilla revival plans...promising future for vanilla in Tonga.

Table 10. The ordinary run analysis of the distribution of VNPV infected plants along rows of vanilla plants in Plot 1.

Rows	U	m	N	E(U)	s(U)	Z
1	7	3	49	6.6	0.736	1.22
2	7	4	49	8.3	0.985	-0.812
3	7	3	50	6.64	0.956	0.899
4	7	4	50	8.36	0.977	-0.880
5	5	7	50	13.04	1.64	-4.59*
6	7	3	50	6.64	0.73	1.17
7	3	2	49	4.83	0.47	-2.83*
8	7	3	50	6.64	0.73	1.17
9	5	2	49	4.8	0.475	1.47
10	5	2	50	4.84	0.471	1.40
11	11	6	49	11.5	1.44	0
12	11	5	50	10	1.212	1.237
13	9	8	50	14.44	1.846	-2.68*
14	9	5	50	10	1.21	-0.413
15	9	5	48	9.9	1.23	-0.325
16	9	4	50	8.36	0.448	2.54

* - significant at P = 0.05

Table 11. Doublet analysis of the distribution of VNPV infected plants along vanilla rows in Plot 3.

Rows	D	m	N	E(D)	s(D)	Z
1	1	9	49	1.46	1.18	-0.389
2	3	11	50	2.2	1.45	0.55
3	6	21	50	8.4	2.83	-0.848
4	4	16	49	4.89	2.21	-0.4
5	7	18	35	8.74	2.87	-0.606
6	3	14	48	3.79	1.9	-0.415
7	0	7	45	0.93	0.89	-1.04
8	5	18	49	6.2	2.44	-0.491
9	4	13	50	3.12	1.73	0.51
10	2	9	35	2.05	1.39	-0.036
11	2	12	48	2.75	1.62	-0.46
12	0	4	49	0.24	0.48	-0.5
13	2	10	50	1.8	1.31	0.625
14	2	7	48	0.875	0.916	1.228
15	5	12	47	2.8	1.64	1.34
16	1	11	49	2.2	1.468	-0.817
17	8	18	49	6.2	2.44	0.737
18	6	15	50	4.2	2	0.9
19	5	20	50	7.6	2.7	-0.96
20	3	12	41	32	1.75	-0.114

FAO STATISTICS, 2009....WORLD CURED VANILLA BEAN PRODUCTION....

Madagascar...3500 tonnes, Indonesia....3400t, China....1350t, Papua New Guinea....400t, Mexico....390t, Turkey....290t, Tonga....202t, Uganda....170t, French Polynesia....60t, Comoros....42t.

Table 12. Ordinary run analysis of the distribution of VNPV infected plants along the vanilla rows in Plot 3.

Rows	U	m	N	E(U)	s(U)	Z
1	16	9	49	15.7	2.04	0.392
2	16	11	50	18.16	2.37	-0.7
3	30	21	50	25.36	3.4	1.5
4	23	16	49	22.55	3.03	0.31
5	20	18	35	18.4	2.9	0.724
6	22	14	48	20.8	2.81	0.6
7	12	7	45	12.8	1.7	-0.17
8	26	18	49	23.7	3.21	0.872
9	19	13	50	20.24	2.67	-0.277
10	14	9	35	14.37	2.2	0.059
11	20	12	48	19	2.55	0.588
12	9	4	49	7.3	0.98	2.24
13	16	10	50	16	2.21	0.226
14	10	7	48	11.95	1.67	-0.868
15	15	12	47	17.87	2.56	-0.925
16	19	11	49	18.06	3.03	0.475
17	21	18	49	23.77	3.21	-0.7
18	19	15	50	22	2.92	-0.856
19	30	12	50	24	3.35	1.94
20	18	12	41	17.97	2.6	0.2

A Fair Go for Tongan Vanilla Growers..... Queen Fine Foods, New Zealand's largest distributor of vanilla products used widely in Kiwi homes, has entered into a partnership with the people of Tonga to reinvigorate their vanilla industry. The **Queen Fine Foods** initiative works with growers to develop sustainable and organic farming practices. It teaches farmers not only how to grow high quality beans, but to cure their crop and add value. Growers who join the partnership receive a long term supply agreement with Queen, which guarantees certainty of income for years to come....press release by Food Marketing Ltd...online...

CHAPTER 6. SURVEY OF THE VANILLA PLANTATIONS OF TONGATAPU AND VAVA'U.

General distribution, source of infection and spread of VNPV in Tonga.

Two hundred and thirteen vanilla plantations were selected from the MAFF list of vanilla growers using the first 3 digits of a random numbers table. Forty-five were finally surveyed due to the large number of neglected plantations. Plantations less than 2 years old were also surveyed to find infected vanilla shoots and determine presence of symptoms on the cuttings. This would indicate spread of the VNPV through infected planting material (Pone, 1988).

Scoop.co.nz...Vanilla Rehab in Tonga.

The new Queen Fine Foods Tongan vanilla rehabilitation initiative that is working with local grower families has been inspected by the King of Tonga, Tupou VI, who recently visited the island group of Vava'u. **New Tongan Vanilla Programme Gains Momentum.**

TABLE 13. VNPV INFECTION AND THE INCIDENCE IN EACH PLANTATION SELECTED FROM VAVA'U ISLAND.

Plot Number	Sample	Symptomatic	Percentage
V2	111	16	14
V3	94	0	0
V4	87	0	0
V5	74	26	35
V6	87	36	22
V7	164	36	22
V8	321	49	15
V9	148	8	5
V11	87	15	17
V12	91	7	8
V13	142	0	0
V14	252	16	6
V15	97	55	57
V17	112	5	4
V18	130	108	83
V19	124	6	5
V20	173	48	28
V21	127	14	11
V22	99	2	2
V23	83	5	6

Source - Pone, 1988.

Table 14. VNPV INFECTION AND THE INCIDENCE IN EACH PLANTATION SELECTED FROM TONGATAPU ISLAND.

Plot Number	Sample	Symptomatic	Percentage
T1	160	21	13
T2	129	1	1
T4	109	5	5
T5	152	2	1
T6	140	7	5
T7	136	5	4
T8	90	1	1
T9	160	10	6
T10	75	6	8
T11	80	1	1
T12	147	5	3
T13	281	6	2
T14	24	5	21
T15	115	24	21
T16	55	0	0
T19	114	7	6
T21	147	12	8
T23	85	23	27

Source - Pone, 1988.

Infected vanilla plants grown from infected cuttings.

Table 15. Infected vanilla plants growing from symptomatic cuttings.			
Plot number	sample	VNPV shoots	VNPV cuttings
T3	264	17	17
T17	55	4	4
T18	97	0	0
T20	137	1	1
V1	106	4	4
V10	212	32	32
V29	45	0	0
V30*	592	295	295

*-all plants in this plantation was examined due to the large number of infected cuttings.
Source: Pone, 1988.

General distribution of VNPV in Tonga.

The survey of the plantations on the islands of Tongatapu and Vava'u Islands showed that VNPV symptoms are widespread on plantations in these islands. Most of the plantations surveyed have low incidence at less than 30%. A few (7%) had high VNPV incidence of greater than 30% of the plants surveyed. The age of the plantations surveyed were between 3 and 8 years old.

The variability of the VNPV incidence suggest that the virus is not spreading quickly. Although in isolated cases it may spread quickly, when certain conditions like warm temperatures and presence of inoculum and vectors as in the case of Plot 1.

"The infection of isolated plants by viruses is not normally of economic importance. Viruses become important when they cause epidemics. Epidemics maybe defined as a change in disease intensity with time. Disease intensity has 2 components; 1. disease incidence, which is the proportion of infected plants 2. disease severity, which is the number of plants infected (Madden and Campbell, 1986)".

The number of infected plants and plantations in Tonga during this study was large enough to be considered an epidemic. Of the 47 plantations surveyed, 41 plantations had VNPV symptomatic plants along the rows surveyed. That is 87% of all plantations surveyed. The incidence of VNPV ranged from 1 to 83%.

The rapid spread of VNPV in Plot 1 where the VNPV infected plants jumped from 25 plants in August to 66 in December, suggest that if conditions are right VNPV can spread rapidly and destroy plantations quickly.

The 100% VNPV symptomatic cuttings giving rise to symptomatic plants (Table 15) suggest that the cuttings is the most important agent of dispersal over long distances and between the islands.

Fairgo for Tongan Vanilla... this initiative will deliver New Zealanders a better quality final product. "Most home bakers probably don't think about what it takes to make that little aromatic bottle of Queen vanilla extract they have in their pantry. Our passion is for vanilla and vanilla growers, so partnerships where we work hand in hand with growers mean we can manage the product quality from farm to pantry," he said. "Together with Queen's 100-year-old exclusive extraction technique, this ensures we are producing the world's finest vanilla products. For our New Zealand customers it means they get unsurpassed flavour in their home baking."Queen has supported vanilla farmers in a number of countries and produces many premium quality vanilla products available nationwide through participating supermarkets and **specialty food stores.**

Food Marketing Ltd....press release...online

CHAPTER 7. GENERAL DISCUSSIONS ON VNPV DISTRIBUTION.

The allocation of funds and manpower for the control of a disease depends on how widespread the problem is. The larger the area affected, the more money and manpower required for virus control, since most virus control methods with vegetatively propagated plants will largely depend on removal of infected plants.

"The surveys show that VNPV is widespread on the main growing islands of Vava'u and Tongatapu, with mostly low incidence but can be very high in some isolated cases. The most likely source of infection are cuttings which then spread to other plants in the plantation through mechanical means and vectors. The rate of spread depends on the presence of aphids and transmission of infected sap between plants, as suggested by successful aphid and mechanical transmission of the virus".

Since VNPV was found to be widespread on both the main growing vanilla areas of

Tongatapu and Vava'u Islands, the only solution was the removal of all infected plants showing symptoms of the virus. Two thousand leaflets printed in the Tongan language were distributed to farmers with advise on how to control the disease (Pone and Pearson, 1989). Both advisory staff and farmer trainings were carried out in Tongatapu and Vava'u. All symptomatic vanilla plants on all plantations on Tongatapu and Vava'u were to be removed and destroyed and replaced with healthy cuttings. It was interesting to note that the total cured vanilla export jumped from 20 tonnes in 1987 to 70 tonnes about 3-4 years later following the application of the VNPV control   strategies.

Two dimensional distribution of VNPV in vanilla plots 1, 2 and 3.

The comparison of the mean and variance of the 3 plots found the distribution of the infected plants to follow a negative binomial or contagious distribution. The variance was greater than the mean number of infected plants in the quadrats for each of the plots studied.

For the smaller quadrat size, the test for significant deviations from the Poisson distribution found Plot 1 to differ significantly (P = 0.001). This means the highly contagious distribution was effected by a vector. This plot had a large number of aphids resident on a mat of *Emilia sonchifolia* and *Sonchus oleraceous* weeds covering the whole plantation. Plots 2 and 3 did not differ significantly at this level of significance.

When the size of the quadrats were increased to twice the size of the small quadrats, all the Plots differed significantly from the Poisson distribution at P = 0.10. Only plots 1 and 3 differed significantly from the Poisson distribution at P = 0.05 and P = 0.025. This suggest that a vector is spreading the virus within the plantations as proposed by Hollings and Brunt, 1981a and Matthews 1982...**"viruses do not spread on their own, but are carried by vectors. The vector for potyviruses are usually aphids transmitting** the virus in a non-persistent manner ….furthermore Madden and Campbell, 1986... proposed…"the spatial pattern of infected plants will be clustered or clumped if the infection of a given plant increases the probability of another plant nearby being infected. In short, with clustering, infected plants are, on average, grouped together and healthy plants are grouped".

Linear distribution of VNPV infected plants in vanilla rows.

Plot 1 had significant clustering (P = 0.05) in rows 5, 7 and 13 in both doublet and ordinary run analyses. The disease intensity in Plot 1 was already very high with the number of infected plants and vectors present. The number of diseased plants had jumped from 25 to 66 during the 5 months when the plot was monitored, an increase of 160%. The evidence from the aphid transmission study gives us some confidence with our conclusion that the presence of the large number of aphids in Plot 1 is conducive to the rapid spread and clustering of infected vanilla plants. Firstly, aphids are known to visit vanilla plants and feed on their leaves. Secondly, aphids are known to transmit VNPV between *N.benthamiana* plants. The warm temperatures between September and December is not only perfect for the weed hosts and aphid reproduction but the Thermal Inactivation Point study suggest that VNPV may be more infective and higher temperatures. Tonga has a cool period from around April to August and warm period from September to March.

"It is known that the number of aphids present in a crop infected with an aphid transmitted virus often correlates to the incidence of infected plants (Thresh, 1986). This type of information has been used, together with weather information, to forecast disease outbreaks from the weather patterns that affect aphid population growth (Heathcote, 1986)".

The positive test result also suggest there may be an effect of cultural practices along the rows. Looping and flower initiation requires the grower to cut off some vanilla leaves and shoots and perhaps the infected sap is transferred along the row. Growers were advised to use bleach or a lighter to sterilise tools while working in the plantation to ensure infected sap are not transferred between plants.

It should be noted that during the 5 months period of VNPV mapping and monitoring of Plots 1 and 2, no herbaceous weeds or aphids were found in Plot 2. There were only 2 new infected plants in that plot during the 5 months.

No significant clustering was found in the vanilla rows in Plot 3, although it appears to be the case visually. There were some rows (14, 15, 18) with some very high positives values in the doublet analysis and high negative values in the ordinary analysis, but were much lower than required by the test. This means that there is some clustering in Plot 3, although not significant at this stage.

It is noted from other studies such as the epidemiology of Cucumber mosaic cucumovirus (CMV) on kava (*Piper methysticum*) (Davis, R and Brown, J, (1996) , which is another virus transmitted dieback disease of a Pacific crop ,that CMV infected plants are clustered at the beginning of the epidemic, then becomes random as most of the plants are infected. As Davis and Brown (1996) puts it, "plants were showing a decrease in aggregation" after 24-25 weeks of monitoring, using Lloyd's Patchiness Index.

The epidemics usually starts with an infected plant that infects nearby plants causing a clustered distribution. This may depend on the presence of vectors as evidenced by the differences in the number of infected plants between Plots 1 (25) and 2 (2) during the 5 months of mapping. Similar patterns are also seen in kava plantations in Tonga (Davis and Brown, 1996).

Vanilla in Tonga....ISLANDS BUSINESS...It is one of 28 enterprises supported by the European Union-funded Increasing Agricultural Commodity Trade (IACT) Project, implemented by the Land Resources Division of the Secretariat of the Pacific Community (SPC). After a successful product display at the World Food and Beverage Expo 2013 in Tokyo in April, Heilala scored a new market for its brands of vanilla-based products, adding Japan to its list of markets, which already include New Zealand, Australia, Malaysia, Singapore and the United States. Heilala Vanilla, owned by Jennifer Boggiss and her father John Ross, began operation in 2002 as an aid project after a cyclone destroyed a local village.

CHAPTER 8. CONCLUSIONS

Numerous pathogens have been reported to attack vanilla in Tonga, but the shoot necrotic lesions, diffuse sunken chlorotic streaks on leaves and vine dieback have been conclusively proved to be caused by the Vanilla Necrosis Potyvirus as characterised and reported by Pone, 1988; Pearson and Pone, 1988; Pearson *et al*, 1990. There are many other studies of this virus as well.

The relative incidence of the VNPV on *Vanilla fragrans* in Tonga appear to be much higher than similar studies on *Vanilla tahitiensis* in French Polynesia (Wisler *et al*, 1987a,b).

VNPV is widespread on the main vanilla growing islands of Tongatapu and Vava'u with low incidence in most plantations and some isolated cases of high VNPV infection. Infected cuttings used for new plantations is the main source of inoculum in new plantations. No alternative hosts of VNPV in

Tonga has been found.

Both spatial and linear distribution of VNPV are clumped or clustered according to the quadrat, doublet and ordinary run analyses. It is an indication that vectors such as aphids are spreading the virus within the plantation. Clumping of infected plants along rows suggest that growers may also carry the virus on their tools as they work along the rows.

Control strategies were established with advisory staff and farmer training to include all the new knowledge from this study. It was noted that the export of cured vanilla beans increased from 20 tonnes to 70 tonnes in the later years. It may have been due to better disease control or increased acreage or both.
It conclusively proves also that the VNPV epidemic in Tonga was effectively halted and reversed. Vanilla growers in Tonga can continue to earn a living from this valuable and popular flavoring in the future.

Literature Cited.

1. Broadbent, L (1964). Control of Plant Viruses, p 330-364. In *Plant Virology*. Eds. Corbett, M.K. and Sisler, H.D., University of Florida Press, 527 pp.

2. Davis, R.I. And Brown, J.F. (1996). Epidemiology and Management of Kava Dieback Caused by Cucumber Mosaic Virus. *Plant Disease*. August. 917-921.

3. Edwardson, J.R. and Zettler, F.W. (1986). Odontoglossum Ringspot Virus. pp 233-247. In *The Plant Viruses*, Eds. M.H.V. van Regenmortel and H. Fraenkel-Conrat. Plenum Press, New York. 424 pp.

4. Elliott, J.M. (1977). Some methods for the Statistical Analysis of Benthic Invertebrates. Second Edition. *Freshwater Biological Association Scientific Publications* No. 25. United Kingdom. 160 pp.

5. Francki, R.I.B. (1970). Cymbidium Mosaic Virus. *CMI/AAB Description of plant viruses*, Number 27, 3pp.

6. Hamilton, R.I., Edwardson, J.R., Francki, R.I.B., Hsu, H.T., Hull, R., Koenig, R. and Milne, R.G. (1981). Guidelines for the identification and characterisation of plant viruses. *Journal of General Virology* 54, 223-241.

7. Heathcote, G.D. (1986). Virus yellows of sugar beet. pp 399-417. In *Plant Virus Epidemics, Monitoring, Modelling and Predicting Outbreaks*. Eds. McClean, G.D., Garret, R.G. and Ruesink, W.G. Academic Press, Sydney. 550 pp.

8. Hollings, M. and Brunt, A. (1981a). Potyvirus Group. *CMI/AAB Descriptions of Plant Viruses*. No.245.7pp.

9. Jensen, D. D., (1951). Mosaic or Black Streak Disease of Cymbidium Orchids. *Phytopathology* 41, 401-413.

10. Jensen, D.D. and Gold, A. H. (1955). Hosts, Transmission and Electron Microscopy of Cymbidium Mosaic Virus with special reference to Cattleya Leaf Necrosis. *Phytopathology* 45, 327-334.

11. Jensen, D.D. and Gold, A.H. (1951). A Ringspot Virus of Odontoglossum Orchids. Symptoms, Transmission and Electron Microscopy. *Phytopathology* 41, 401-413.

12. Kado, C.I. and Jensen, D. D. (1964). Cymbidium Mosaic Virus in Phalaenopsis. *Phytopathology* 54, 975-977.

13. Leseman, D.E. and Koenig, R. (1985). Identification of Bean Yellow Mosaic in Masdevallia. *Acta Horticulturae* 164, 347-354.

14. Leseman, D.E. and Vetten, J.J. (1985). The occurrence of tobacco rattle and turnip mosaic viruses in Orchid spp. and an unidentified potyvirus in Cyprideum calceous. *Acta Horticulturae* 164, 45-54.

15. Liefting, L., Pearson, M.N. and Pone, S.P. (1992) The Isolation and Evaluation of Two Naturally Ocurring Mild Strains of Vanilla Necrosis Potyvirus for control by Cross Protection. *Journal of Phytopathology* Vol 136, Issue No.1. 9-15. September, 1992.

16.   Madden, L.V. and Campbell, C.L. (1986). Descriptions of Virus Diseases in Time and Space. pp 273-293. In *Plant Virus Epidemics, Monitoring, Modelling and Predicting Outbreaks*. Eds. McClean, G.D., Garrett, R.G. and Ruesink, W.G. Academic Press, Sydney. 550 pp.

17. Madden, L.V., Louie, R., Abt, J.J. and Knoke, J.J. (1982). Evaluation of Tests for Randomness of Plants. *Phytopathology* 72, 195-198.

18. Matthews, R.E.F. (1982). Classification and nomenclature of Viruses. Fourth Report of the International Committee on Taxonomy of Viruses. *Intervirology* 17, 1-200.

19. Paul, H.L. (1975). Odontoglossum Ringspot Virus. *CMI/AAB Descriptions of Plant Viruses*. No. 155. 4pp.

20. Pearson, M.N. and Cole, J.S. (1986). The effects of Cymbidium Mosaic Virus and Odontoglossum Ringspot Virus on the growth of *Cymbidium* orchids. *Journal of Phytopathology* 117, 193-197.

21. Pearson, M.N. and Pone, S.P. (1988) Viruses of vanilla in the Kingdom of Tonga. *Australasian Plant Pathology* 17 (3), 59-60.

22. Pearson, M.N., Brunt, A.A. and Pone, S.P. (1990). Some hosts and properties of a potyvirus from *Vanilla fragrans* in the Kingdom of Tonga. *Journal of Phytopathology* 128:46-54.

23. Pone, S.P. (1988) An investigation of 3 virus diseases of *Vanilla fragrans* (Salisb.)Ames in the Kingdom of Tonga. *MSc Thesis*, University of Auckland, New Zealand. 157 pp.

24. Pone, S.P. (2013). *Plant Protection in the Pacific*. Second Edition. Rainbow Enterprises, Auckland. 114 pp.

25. Pone, S.P. (1989). (unpublished). Weeds around MAFF *Vanilla* plantations were tested using ELISA for the presence of VNPV. MAFF LAB reports. MAFF Research Division. Tonga.

26. Pone (unpublished). Developing an ELISA test to Vanilla Necrosis Potyvirus in the Kingdom of Tonga.

27. Project Performance Audit Report, Tonga Development Bank, May, 1988; unpublished.

28. Thresh, J.M. (1986). Plant Virus Disease Forecasting, pp 359-386. In *Plant Virus Epidemics, Monitoring, Modelling and Predicting Outbreaks*. Eds. McClean, G.D., Garrett, R.G. and Ruesink, W.G. Academic Press, Sydney. 550 pp.

29. van der Plank, J.E. (1960). Analysis of Epidemics, pp 229-289. In Plant Pathology, Volume III. Eds. Horsfall, J.G. and Dimond, A.E., Academic Press, New York, 675 pp.

30. Walkey, D.G.A. (1985). *Applied Plant Virology*. Heinemann, London. pp 329.

31. Wisler, G.C., Zettler, F.W. and Mu L. (1987a). Viruses affecting *Vanilla* in French Polynesia. *American Orchid Society Bulletin* 56 (4), 1987.

32. Wisler, G.C., Zettler, F.W. and Mu L. (1987b). Virus infections of *Vanilla* and other orchids in French Polynesia. *Plant Disease* 71 (12), 1125-1129.

Figure 4. Map of the Vanilla Plantations surveyed on Vava'u Island, Kingdom of Tonga.

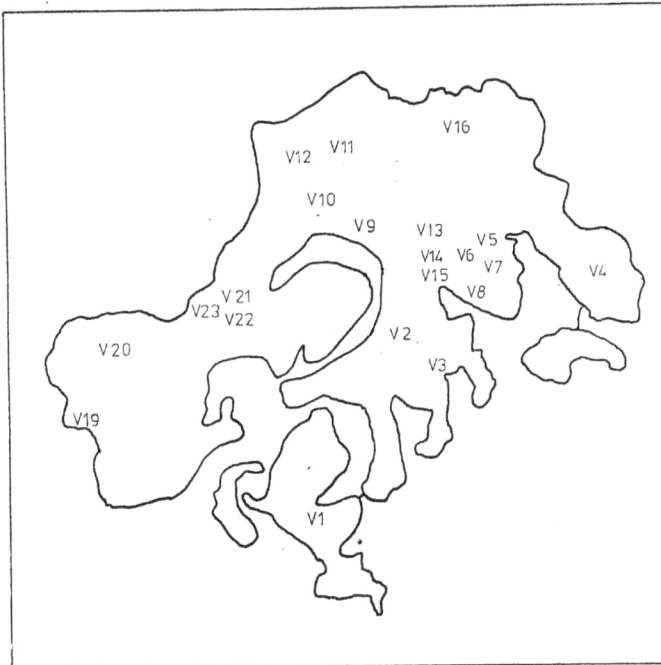

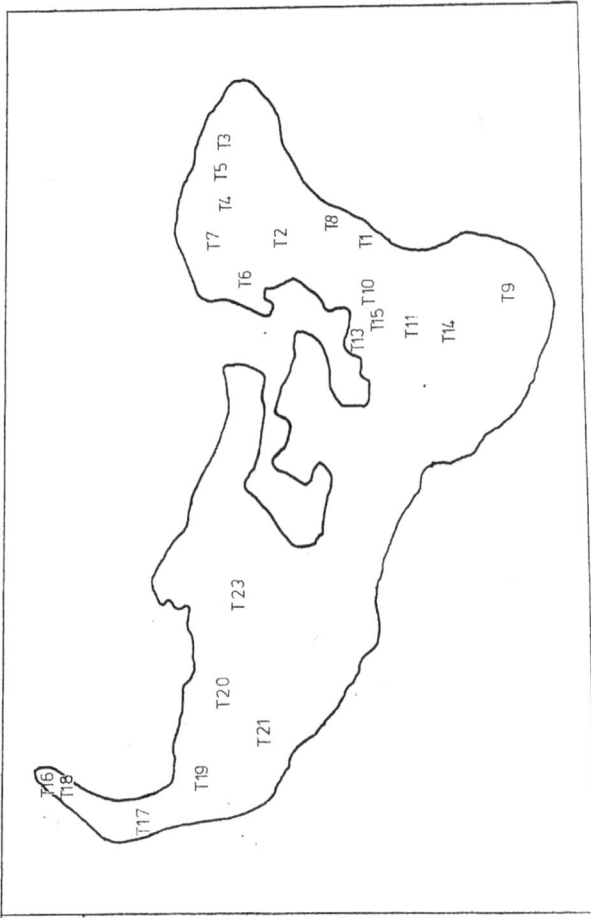

FIGURE 5. Map of the Vanilla Plantations Surveyed on Tongatapu Island, Kingdom of Tonga.

www.ingramcontent.com/pod-product-compliance
Lightning Source LLC
Chambersburg PA
CBHW060644210326
41520CB00010B/1729